WONDER
STARTERS

SO-ASI-534

Night

Pictures by JANET AHLBERG

Published by WONDER BOOKS
A Division of Grosset & Dunlap, Inc.
A NATIONAL GENERAL COMPANY
51 Madison Avenue   New York, N.Y. 10010

Published in the United States by Wonder Books, a Division
of Grosset & Dunlap, Inc., a National General Company.

ISBN: 0-448-09670-6 (Trade Edition)
ISBN: 0-448-06390-5 (Library Edition)

FIRST PRINTING 1973

© Macdonald and Company (Publishers), Limited, 1972, London.
All rights reserved.

Printed and bound in the United States.

Library of Congress Catalog Card Number: 73-1973

It is time for me to go to bed.
Soon it will be dark.

1

Most people sleep at night.
Many animals sleep at night, too.

2

Some people work at night.
They sleep in the daytime.
This nurse works at night.

3

This man is a night watchman.
He is looking after a factory.
He looks out for robbers.

4

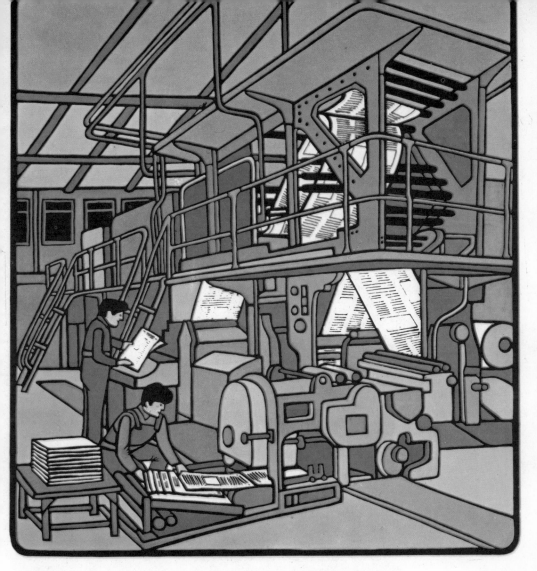

This machine prints newspapers.
Men work the machine at night.
People can buy the newspapers
in the morning.

Some people travel at night.
These people are sleeping
on a train.
They sleep while the train goes along.

6

Airplanes often fly at night.
People sleep in the seats.

7

Ships often sail all night.
They carry special lights.
Sailors on other ships see them.

8

The lighthouse stands on an island.
The lighthouse shows
where the rocks are.
Its light shines at night.

Night happens like this.
The sun shines on the world.
It shines on one side.
Where the sun is shining, it is day.

10

On the other side of the world,
it is night.

The world is always turning.
So night comes
to the places where it was day.
12

Night begins in the evening.
Night ends in the morning.

The moon shines at night.
People can see by moonlight.

14

Stars shine at night, too.
They do not give much light.

cat

fox

badger

mouse

Many birds and animals
can see in the dark.

16

Owls hunt at night. They fly about,
looking for little animals.
Owls can see well in the dark.

These animals are bats.
They fly at night.
Many bats eat insects.
18

In some countries
the bats are huge.
People call bats like these
flying foxes.

Many insects fly at night.
They often fly into lights.
20

eyes

feelers

legs

wing cases

These insects are fireflies.
They can shine at night.

See for yourself.

Make a model of the world.

Shine a flashlight on the model.

Pretend the light is the sun.

You can see how day and night happen.

22

Starter's **Night** words

bed
(page 1)

factory
(page 4)

people
(page 2)

robber
(page 4)

animal
(page 2)

print
(page 5)

night watchman
(page 4)

newspaper
(page 5)

23

buy
(page 5)

ship
(page 8)

train
(page 6)

light
(page 8)

airplane
(page 7)

sailor
(page 8)

seat
(page 7)

lighthouse
(page 9)

rocks
(page 9)

bat
(page 18)

world
(page 10)

insect
(page 18)

stars
(page 15)

flying fox
(page 19)

owl
(page 17)

flashlight
(page 22)

25